Habitat Days and Nights

DAY AND NIGHT ON THE Prairie

by Ellen Labrecque

PEBBLE
a capstone imprint

Published by Pebble, an imprint of Capstone.
1710 Roe Crest Drive, North Mankato, Minnesota 56003
capstonepub.com

Copyright © 2022 by Pebble, a Capstone imprint.
All rights reserved. No part of this publication may be reproduced in whole or in part, or stored in a retrieval system, or transmitted in any form or by any means, electronic, mechanical, photocopying, recording, or otherwise, without written permission of the publisher.

Library of Congress Cataloging-in-Publication Data is on file with the Library of Congress.
ISBN: 9781663976925 (hardcover)
ISBN: 9781666327953 (paperback)
ISBN: 9781666327960 (ebook PDF)

Summary: Spend a day and night on the prairie! Learn about this grassy habitat through the exciting animals that call it home. Spot prairie dogs popping aboveground as morning sun floods a field. Join giant bison as they graze. Hunt with a burrowing owl as the sun sets. After dark, stake out prey with a coyote in dense grass. What will tomorrow bring on the prairie?

Image Credits
Flickr: Bert Cash, 19; iStockphoto: debibishop, Cover (prairie), 1, Hailshadow, Cover (ground squirrel), 1, NetaDegany, 15; Mighty Media, Inc.: 20, 21; Shutterstock: Annette Shaff, 13, Anton Foltin, 9, Bob Pool, 5, Edwin Butter, 6, Greg Birkett, 11, Jim Nelson, 18, Liz Weber, 14, 16, Michael Dante Salazar, 12, Peter Kirillov, 7, vagabond54, 8, Viktor Loki, 17

Editorial Credits
Jessica Rusick, editor, media researcher; Kelly Doudna, designer, production specialist

All internet sites appearing in back matter were available and accurate when this book was sent to press.

Table of Contents

What Is a Prairie? 4
Morning 6
Noon 8
Late Afternoon 10
Evening 12
Night 14
Late Night 16
Dawn 18
    Prairie Dog Activity 20
    Glossary 22
    Read More 23
    Internet Sites 23
    Index 24
    About the Author 24

Words in **bold** are in the glossary.

What Is a Prairie?

A prairie is a flat and grassy **habitat**. Prairies are found in North America. They are covered in grasses and other plants. There are few trees.

Many animals live on prairies. Some come out during the day. Others come out at night.

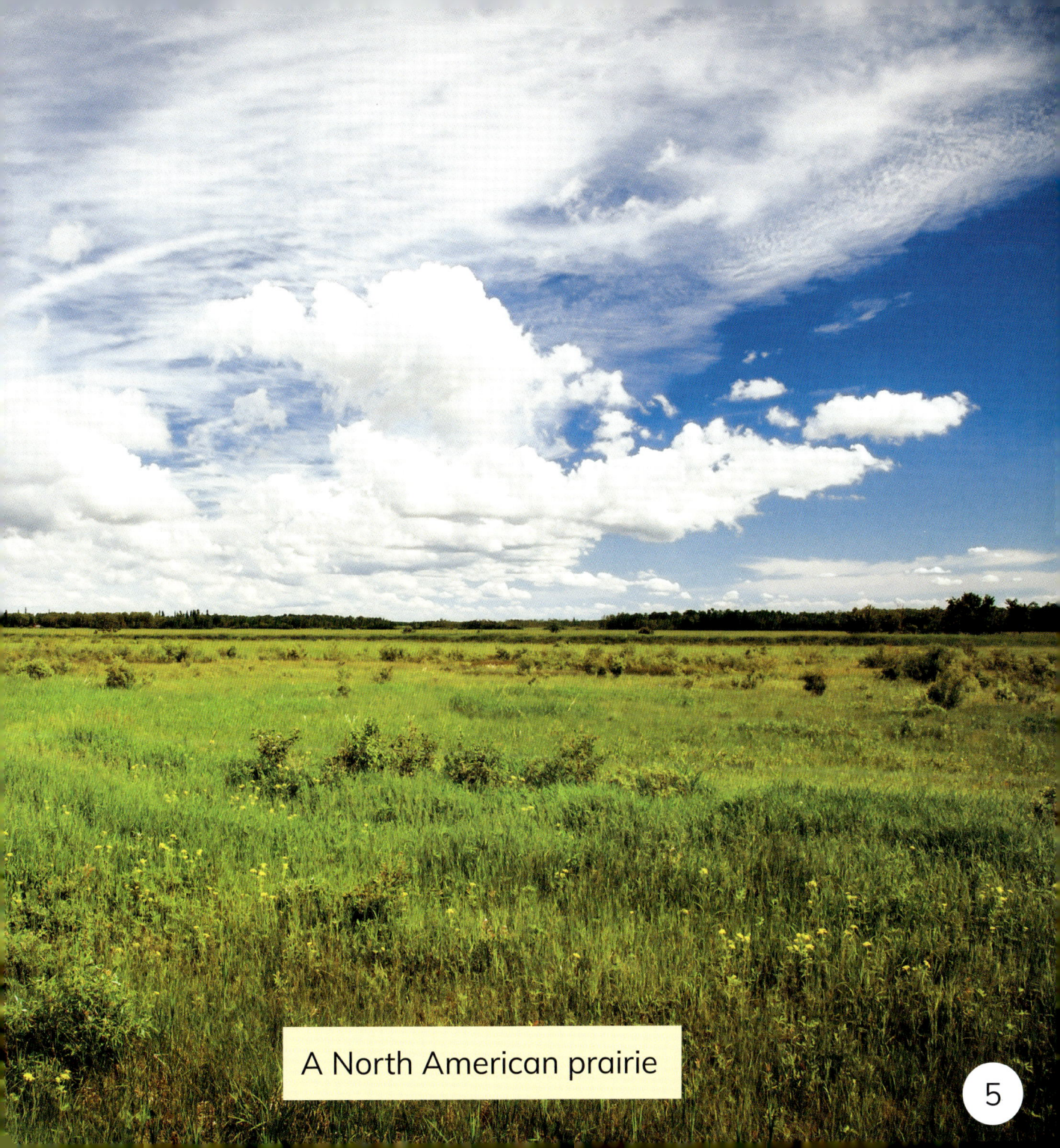
A North American prairie

Morning

The sun rises. Prairie dogs pop out of their **burrows**. The prairie dogs start gathering food. They find grass and seeds to eat.

Some prairie dogs stand on their back legs. They are watching for **predators**. The prairie dogs spot a hawk. They yip loudly to warn others of the danger.

Noon

The air warms. A long-billed curlew sits on its nest. The nest is a hole lined with grass. It is hidden. This keeps it safe from predators.

Long-billed curlew

Bison

A bison herd eats grass nearby. Each bison will eat about 24 pounds (11 kilograms) of food today! Bison are the largest **mammals** in North America.

Late Afternoon

A pronghorn herd eats **shrubs**. The herd also watches for predators. Pronghorns have strong eyesight. They can see things more than 2 miles (3 kilometers) away!

A predator comes near. The pronghorns sprint away. Pronghorns can run at 60 miles (97 km) per hour!

Pronghorns

Evening

The sun is setting. A burrowing owl runs along the ground. It grabs a ground squirrel in its **talons**. The owl brings the animal back to its burrow. The owl stores the squirrel to eat later.

Ground squirrel

Burrowing owl

Night

Stars twinkle overhead. A porcupine eats leaves and flowers. The porcupine is covered in sharp hairs called quills. These can hurt predators.

Porcupine

Coyote

A coyote roams nearby. Coyotes can smell and hear well. This helps them hunt **prey**. The porcupine shakes its quills at the coyote. This is a warning to stay back.

Late Night

A skunk searches for food. Its white fur warns nighttime predators to stay away. Predators sometimes come too close. The skunk sprays them with smelly liquid!

Skunk

Night snake

A night snake slithers out from between rocks. Its forked tongue moves in and out. The snake smells with its tongue. It searches for frogs and lizards to eat.

Dawn

A mole digs a tunnel underground. It touches its nose to the dirt. Moles use touch to sense where they are.

Blackbirds chirp as the sun rises. The sun warms their feathers. Another day on the prairie has begun.

Blackbird

Mole

Prairie Dog Activity

What You Need:

- pencil
- brown construction paper
- scissors
- black marker
- glue
- popsicle stick
- paper cup
- green marker

What You Do:

1. Draw a prairie dog on a sheet of brown construction paper.

2. Cut out the prairie dog. Draw a nose, mouth, and other features with a black marker.

3. Glue the prairie dog to one end of a popsicle stick.

4. Cover the outside of a paper cup in brown construction paper. Color the cup's rim green.

5. Cut a slit in the bottom of the cup.

6. Slide the popsicle stick through the slit.

7. Slide the stick up and down to make the prairie dog pop out of its burrow!

Glossary

burrow (BUHR-oh)—a tunnel or hole in the ground made or used by an animal

habitat (HAB-uh-tat)—the natural place and conditions in which a plant or animal lives

mammal (MAM-uhl)—a warm-blooded animal that breathes air; mammals have hair or fur; female mammals feed milk to their young

predator (PRED-uh-tur)—an animal that hunts other animals for food

prey (PRAY)—an animal hunted by another animal for food

shrub (SHRUHB)—a plant or bush with woody stems that branch out near the ground

talon (TAL-uhn)—a long, sharp claw

Read More

Gardeski, Christina Mia. *All About Grasslands.* North Mankato, MN: Capstone, 2018.

Loria, Laura. *What Are Grasslands?* New York: Britannica Educational Publishing, 2019.

Schuh, Mari. *Prairie Dogs.* North Mankato, MN: Capstone, 2020.

Internet Sites

Britannica Kids—Grassland
kids.britannica.com/kids/article/grassland/346127

National Geographic Kids—Prairie Dog
kids.nationalgeographic.com/animals/mammals/facts/prairie-dog

PBS Kids—Nature Cat: The Prairie Is Awesome!
pbskids.org/video/nature-cat/3030744266

Index

bison, 9
blackbirds, 18
burrowing owls, 12
burrows, 6, 12

coyotes, 15

grasses, 4, 6, 8, 9
ground squirrels, 12

hawks, 7

long-billed curlews, 8

moles, 18

night snakes, 17
North America, 4, 9

porcupines, 14, 15
prairie dogs, 6, 7
pronghorns, 10

skunks, 16

About the Author

Ellen Labrecque is the author of more than 100 nonfiction children's books. She lives in Bucks County, Pennsylvania, with her husband and two kids. She has the best writing partner in the world—her dog, Oscar. An avid reader and runner, Ellen is a morning person. On most days, she is up before the sun.

JAN 07 2022